# Trade Economic Analysis

Dr. Mieszko Kaczorowski

# DEDICATION

To My Wonderful Readers

# CONTENTS

# ACKNOWLEDGMENTS

Thank you to everyone who strives to grow and help others grow

# 1 ABSTRACT

The law of comparative advantage is the cornerstone of international trade. Economic growth, agriculture, poverty reduction and trade have a very complicated relationship. Trade analysis helps any countryto discover profitable opportunities; hence the study has undertaken with these objectives.To analyze agricultural trade growth, responsiveness, stability and estimate the effects on production, exports, imports and poverty and suggestpolices for boosting Trade.The major agricultural commodities from 1990-91 to 2015-16 were selected. Estimated CAGR, Instability Index, Export import price elasticity's and regression.During the period 1990-91 to 2015-16, wheat export price and rice import price growth rates were found to be more than other cereals. The imports price growth rate of all pulses were more than exports price growth rates indicating India is importing at higher prices. Among the oilseeds crops, mustard import price (11.05 %) and sunflower export price (7.85 %) growth rates are more. The export import quantity growth rate of all cereals, pulses, and oilseeds were found to be positive except rice imports, wheat exports and soybean imports which were negative. Export import prices of all cereals (except maize) and all pulses were found to be unstable. Import price of all oilseed crops were found to be stable (except mustard and sesamum). The study found that among cereals, pulses and oilseeds, rice (1.24 %), peas (2.36 %) and mustard (0.97 %) have high export elasticities indicating these crops highly responsive to change in international prices. The effect of trade on poverty shows that, production for cereals and pulses, import price for pulses and oilseeds, export price for all the 3 groups have adverse impact on Poverty. Only for oilseeds, production and trade balance have positive and significant effect on poverty.The effect of trade on poverty shows that, export price on all cereals, pulses, oilseeds have adverse impact on poverty. Only for oilseeds, production and trade balance have positive and significant effect on poverty. Thus import prices of cereals, quantity of imports for pulses and export prices of cereals, pulses and oilseeds need to be thoroughly examined and pre-requisite actions as well as policy measures are required. The study advocates that India should design self-reliance policies in pulse production as India is importing pulses at higher prices. India also should design new stable foreign price policy for cereals and pulses as export import prices of all cereals and all pulses were found to be unstable.

Index Terms— International trade, Effects, Productions, Prices, Poverty

# 2 INTRODUCTION

Economic growth, poverty reduction and trade have a very complicated relationship. According to WTO and World Bank, "the key characteristics that have identified as being of relevance to the impact of trade on poverty are: the extreme poor mainly live in rural areas, many earn their livelihoods in the informal sector, the majority lives in a fragile and conflict affected state and women face particular challenges in alleviating poverty (World Bank 2014-15). Top 10 countries (2011)with the largest share of the extreme poor are; 30% India, 10% Nigeria, 5% Cong DEM REP, 3% Ethiopia, 2% Tanzania, 2% Madagascar, 4% Indonesia, 2% Pakistan, 8% China, 6% Bangladesh, 28% rest of the world (World Bank and WTO, 2015).

The law of comparative advantage is the cornerstone of the pure theory of international trade, (Dominick Salvatore, 2004).Almost all Western economists today believe in the desirability of free trade, and this is the philosophy advocated by international institutions such as the World Bank, the I M F, and the WTO. And this was the view after World War II, when Western leaders launched the General Agreement on Tariffs and Trade (GATT) in 1947 (William Krist).As per World Bank Group and WTO, the revolutionary development of global connectivity, digital economy, single window system trade made global trade more promising.

A dramatic increase in developing country participation in trade has coincided with an equally sharp decline in extreme poverty worldwide. Developing countries now constitute 48 percent of world trade, up from 33 percent in 2000, and the number of people living in extreme poverty has been cut in half since 1990, to just under one billion people. Trade has helped increase the number and quality of jobs in developing countries, stimulated economic growth, and driven productivity increases, (The World Bank IBRD IDA).The impact of WTO rules on poverty and development, particularly in developing states argues that current WTO rules are biased against the interests of developing States. It also challenges the orthodox argument that free trade alleviates poverty (Sarah Joseph).

The WTO contributes in more specific ways to food security. Ensuring efficient production and distribution of food supplies is, however, only part of the food security equation.Farmers in developing countries faced declining agricultural prices for almost 50 of the last 60 years, and low prices during the last 20.For much of the period since 1950, many development economists and practitioners have taken the view that low prices for agricultural commodities discourage poverty alleviation in developing countries (Aksoy and Beghin 2004).Many of the poorest people in developing countries depend on agriculture and higher prices can therefore have major implications for poverty reduction (Hertel and Winters 2006).

Much of the trade and development literature has presumed that high food prices are bad for the poor because in low-income countries most poor households, even in rural areas, are net food buyers (Ravallion 1990).Ivanic and Martin (2008) and World Bank (2008), using general equilibrium models that incorporate detailed household-level data, conclude that, worldwide, the 2008 food price hikes may have pushed an additional 100 million people into poverty.

Trade agreements and trade measures are trade-enhancing only at aggregate level: product-specific analyses show that cereal is the only sector that benefits from the joint influence of trade agreements and Sanitary and Phytosanitary Standards (SPSs) bySanteramo.FG, Guerrieri.V and Lamonaca.E (2019).A global simulation on European agricultural liberalization and by comparing it with the Doha simulations infers the poverty effects in the developing world Winters.LA (2005). Trade reform promises to help reduce poverty in rural areas of the developing world (Wiggins.S 2002).

India adopted mixed economy strategy, with self-reliance as the main objective. The market economy depends on price signals to correctly allocate its scarce resources. Economic precision is required on international prices, imports and exports. Sufficient and reliable information is a prerequisite for proper decision making be it domestic market or international marketing. Trade analysis content helps you discover profitable opportunities.

World price means a price for any specific good or service in all countries other than one's own (Financial dictionary, 2012). World price influences on international trade. Under modern capitalism, major commercial exports and imports of a country are conducted in world prices (Encyclopedia, 2010). A country exports goods and services with local price lower than the world price. On other hand, it imports goods and services with higher local prices than the world price (Financial dictionary, 2012).

In the past decade, India emerged as a major agricultural exporter, with exports climbing from just over $5 billion in 2003 to a record of more than $39 billion in 2013 (Alktheeb and Sultan, 2015). India trails behind only the United States in cotton exports. India became the world's seventh-largest exporter of agricultural products in 2013, surpassing Australia (Alktheeb and Sultan, 2015). India emerged as a top bovine meat exporter. India has become a very important player on the global market, especially for rice, cotton, sugar, and beef (buffalo). India has also become a sizeable exporter of soybean meal, guar gum, corn, and wheat, as well as a diverse range of other products.

The inelastic price demand for most agricultural commodities means that lower world prices lead to lower export earnings for developing country exporters. The impact of world commodity price variability on producers and consumers and the effectiveness of price signals in bringing about adjustments in supply and demand depend on the extent to which world market prices are transmitted to domestic markets. (David Hallam, 2003).

Some of the review of literature related to exports imports growth and instability of agricultural commodities are as follows: Glezakos (1973) showed a priori argument that export instability is generally larger in the LDCs than in the DCs and this instability is detrimental to economic growth in the former but not in the latter. Nurkse (1958) and Caine (1958) concluded that export instability is pernicious to economic growth. The export instability could affect foreign exchange earnings and as a result it could have negative impact on imports and economic growth (Chaudhary and Qaisrani, 2002).

Some of the review of literature related to exports market and price signals of India's agricultural commodities are as follows: UNCTAD (2009) reported that India's exports to world are very responsive to income changes. A 1% decline in GDP growth of world will lead to 1.88% decline in India's growth of exports to world (UNCTAD, 2009).Upender (2007) reported that there is a long run as well as short run relationship between export and import price in India. Singh and Sangla (2012) reported that exports from India have experienced instability primarily in the developing countries. Demand for India's exports was found to be relatively elastic to both relative prices and the income of the importing countries. And India has a competitive advantage in a broader range of export commodities. The rapidly increasing world demand for India's exports has played a significant role in the satisfactory exports.

Economic growth, Agricultural production, poverty reduction and trade have a very complicated relationship. Trade analysis content helps the countries to discover profitable opportunities; hence the study has undertaken with the following objectives.

# 3 MATERIALS AND METHODS

## OBJECTIVES

1. To analyze comparative advantage of trade by estimating terms of trade forIndia's Agricultural Commodities
2. To analyze trade growth and behavior by estimating growth rates of India's Agricultural Commodities
3. To analyze trade responsiveness by estimatingexports, imports, price elasticitiesof India's Agricultural Commodities
4. To estimate the effects of international trade on, production, exports, imports and poverty
5. To suggest strategies and polices for boosting trade and poverty reduction

## 1. METHODOLOGY

This is basically foreign trade research. Study Period is 1990-91 to 2016-17. Data on quantity, values, and prices of exports, imports, of cereals, pulses and oilseeds and poverty data were collected. India is trading with 20 to 120 countries. Data sources are FAO Stat, Centre for monitoring on Indian economy (CMIE), National Bank for Agricultural and Rural Development (NABARD). Compound annual growth rates of quantity exports, imports, prices, price elasticities of exports and imports were estimated for cereals, pulses and oilseeds. The compound growth rates, price elasticity's, instability index of trade were estimated using the following formulae's. Regression analysis is performed to capture the effect of trade on production, poverty and trade.

**Growth rate formulae:**

The compound growth rate (r) is be calculated by fitting Exponential function to the variables of interest viz., exports, prices for the period 1990-91to 2014-15.

Yt = Y0 (1+ r )t ----------------------------------------(1)

Assuming multiplicative error term in the equation1, model may be linearized by logarithmic transformation

lnYt = A + Bt + € ----------------------------------------(2)

Where, A(=lnAo) and B(=ln(1+r)) are the parameters to be estimated by ordinary least square regression,

t= time trend in year, r = exp (B) -1

**Price elasticity of exports formulae:**

ΣPe = % change in quantity exports/% change in price.

The percentage change in quantity exports is % ΔQ, and the percentage change in price is % ΔP.

We calculate % ΔQ as ΔQ/Q*ave* and

We Calculate % ΔP as ΔP/P*ave*

*So we calculate the price elasticity of exports as* $(\Delta Q/ Q_{ave})/ (\Delta P/P_{ave})$

**Instability Index Formulae:**

Coefficient of variation = (Standard Deviation/Mean) × 100.

Regression analysis was performed to study the effects of exports on international prices, Quantity of exports, Production, poverty. The results on relevant variables are encouraging with expected sign and magnitudes of coefficients.

The model fitted here is:

$Y=\beta0+\beta1X1+\beta2X2+\beta3X3+\varepsilon$

Where,

Y=dependent variables: Production, Poverty, Total Agri exports, Total Agri imports

Bi (i=0,1,2,3) = Regression coefficients

X1= International Prices

X2= Percentage of Exports

X3= Balance of Agricultural Trade

X4= Percentage of Poverty

ε = Error of the model

3. Results

## 3. RESULTS AND DISCUSSIONS:

### 3.1 Comparative advantage of India's cereals, pulses, and oilseeds

The terms of trade of cereals, pulses and oilseeds are presented in table1. During the period 1990-91 to 2015-16, the terms of trade of wheat, total cereals;pegion pea, gram, lentil, peas (pulses); ground nut, soyabean, mustard, safflower, sunflower, sesamus(oilseeds) were found to be more than 1 indicates India has comparative in these crops in the international markets. The study concludes that except rice, maize and mustard and all cereals, pulses, oilseed crops have comparative advantage during liberalization.

**Table1. The terms of trade of cereals, pulses and oilseeds**

| | | Terms of Trade (%) | | |
|---|---|---|---|---|
| **Major group** | **Crops** | **1990-91 to 2000-01** | **2001-02 to 2015-16** | **1990-91 to 2015-16** |
| **Cereals** | **Rice** | 2.0 | 0.7 | 0.89 |
| | **Wheat** | 0.74 | 1.17 | 1.02 |
| | **Maize** | 4.17 | 0.41 | 0.45 |
| | **Total Cereals** | 1.5 | 1.21 | 1.28 |
| **Pulses:** | **Pegion Pea** | 1.94 | 1.58 | 1.68 |
| | **Gram** | 0.9 | 1.52 | 1.32 |
| | **Lentil** | 1.36 | 1.49 | 1.43 |
| | **Peas** | 2.11 | 1.65 | 1.84 |
| **Oilseeds:** | **Ground Nut** | 3.93 | 1.34 | 1.92 |
| | **Soyabean** | 2.11 | 0.9 | 1.21 |
| | **Niger seeds** | 0 | 0 | 1.92 |
| | **Mustard** | 2.13 | 0.69 | 0.98 |
| | **Saffflower** | 0 | 0 | 6 |
| | **Sesamum** | 2.52 | 1.25 | 1.48 |
| | **Sunflower** | 2.21 | 1.22 | 1.44 |

Source: CMIE Commodities, Accessed on 3rd June 2017

### 3.2. Trade growth and behavior of India's major cereals, pulses and oilseeds

The annual compound growth rates and coefficient of variation of export import quantities and prices of

India's major cereals, pulses, and oilseeds are presented in tables 2, 3,4. Over the course of 26 years period (1990-91 to 2015-16) export price growth rates of cereals (Table2) namely, rice, wheat, maize, total cereals were growing at the rate of -0.80 %, 4.11 %, 0.32 % and 0.34 % and import price growth rates of these cereals were growing at the rate of 3.85 %, 0.28 %, 2.21 % and -0.10 % per annum respectively. It shows that wheat export price and rice import price growth rates were found to be more than other cereals export import prices.

During 1990-91 to 2015-16, the exports and imports price growth rate of all pulses (table 3) were found to be positive, except peas export price growth rate. The imports price growth rate of all pulses were more, compared to exports price growth rate indicating India is importing at higher prices.It is noticed that among the pulses, pigeon pea exports and imports price growth rate was more. During the same period, both exports and imports price growth rate of all oilseeds (table 4) were found to be positive. The export price growth rate of groundnut, nigerseeds, safflower and sunflower were more than imports, while a reverse trend was observed in soybean, mustard and sesamum crops. It is noticed that among the oilseeds crops, mustard import price (11.05 %) and sunflower export price (7.85 %) growth rate was more, compared to other oilseeds export import prices.

During period 1990-91 to 2015-16, the export import quantity growth rate of all cereals, pulses, and oilseeds were found to be positive except rice imports (-14.82 %), wheat exports (-3.05 %) and soybean imports (-1.66 %) quantity growth rate, which were negative. The exports quantity growth rates of rice, total cereals, gram, peas, groundnut, soybean, and mustard were more than the imports quantity growth rates, and the remaining crops have shown a reverse trend quantity growth rates.

The results of coefficient of variation of export import prices and quantity of India's major cereals, pulses, and oilseeds showed that during the period 1990-91 to 2015-16, export import prices of all cereals were unstable except maize import price which witnessed slight stability. The export import quantity of all cereals were found to be stable except export quantity of rice. During the same period, variation in import prices of all pulses were found to be more than export prices, and a similar pattern was observed in imports quantity of all pulses except lentil where export quantity (76.36 %) variation is more, when compared to import quantity (58.54%). It is noticed from the table 4 that export as well as import prices of all pulses were found to be unstable. The exports quantity of all pulses were found to be stable than imports quantity, except lentil where both export as well as import quantity were found to be stable.

During the same period, It is observed from the table 4 that the import price of all oilseed crops were found to be stable except mustard and sesamum import prices. The export quantity of all oilseed crops were found to be unstable except exports quantity of soybean and mustard, whereas imports quantity of all oilseeds crops were found to be stable except imports quantity of mustard.

**Table 2. Growth rate and instability of export-import quantity and price of cereals (1990-2016)**

| Crops | Variables | Quantity growth rate (%) 1990-91 to 2000-01 | Quantity growth rate (%) 2001-02 to 2015-16 | Quantity growth rate (%) 1990-91 to 2015-16 | Price growth rate (%) 1990-91 to 2000-01 | Price growth rate (%) 2001-02 to 2015-16 | Price growth rate (%) 1990-91 to 2015-16 |
|---|---|---|---|---|---|---|---|
| **Rice** | Export | 18.49 | 10.96 | 15.69 | -4.35 | 4.16 | -0.80 |
| | | (116.55) | (171.00) | (123.14) | (355.57) | (198.37) | (204.11) |
| | Import | -13.62 | 20.47 | -14.82 | -1.00 | 0.00 | 3.85 |
| | | (80.21) | (94.69) | (47.71) | (139.93) | (158.27) | (110.71) |
| **Wheat** | Export | 14.69 | -16.13 | -3.05 | 2.35 | 5.61 | 4.11 |
| | | (77.55) | (87.08) | (69.15) | (178.46) | (182.54) | (162.08) |
| | Import | -22.11 | 63.48 | 18.00 | -3.27 | 2.52 | 0.28 |
| | | (95.29) | (48.95) | (53.69) | (197.60) | (129.76) | (146.29) |
| **Maize** | Export | 74.85 | 16.93 | 36.14 | -3.30 | 2.54 | 0.32 |
| | | (77.22) | (139.62) | (80.27) | (114.81) | (473.05) | (160.77) |
| | Import | 106.38 | 28.45 | 45.82 | -2.04 | 2.70 | 2.21 |
| | | (52.83) | (39.05) | (43.28) | (86.48) | (184.93) | (103.07) |
| **Total Cereals** | Export | 9.43 | 3.81 | 8.22 | -3.73 | 6.74 | 0.34 |
| | | (172.36) | (139.62) | (95.70) | (530.86) | (265.62) | (299.96) |
| | Import | -8.70 | 37.40 | 6.83 | -4.21 | 3.88 | -0.10 |
| | | (91.33) | (38.89) | (50.45) | (266.78) | (249.01) | (228.96) |

*Parenthesis in the indicates CV (%)*

**Table 3. Growth rate and instability of export-import quantity and price of pulses (1990-2016)**

| Crops | Variables | Quantity growth rate (%) 1990-91 to 2000-01 | Quantity growth rate (%) 2001-02 to 2015-16 | Quantity growth rate (%) 1990-91 to 2015-16 | Price growth rate (%) 1990-91 to 2000-01 | Price growth rate (%) 2001-02 to 2015-16 | Price growth rate (%) 1990-91 to 2015-16 |
|---|---|---|---|---|---|---|---|
| **Pigeon Pea** | Export | 17.36 | -5.28 | 4.53 | -3.76 | 8.79 | 3.25 |
| | | (251.90) | (77.01) | (85.08) | (672.32) | (165.74) | (192.71) |
| | Import | 7.00 | 1.80 | 12.71 | -0.92 | 9.34 | 4.50 |
| | | (136.70) | (352.93) | (140.88) | (543.63) | (232.07) | (221.69) |

| | | | | | | | |
|---|---|---|---|---|---|---|---|
| **Gram** | Export | 3.03 | 39.79 | 20.11 | -4.54 | 6.81 | 1.44 |
| | | (108.92) | (121.40) | (75.78) | (180.55) | (300.51) | (193.87) |
| | Import | -8.00 | 4.71 | 7.43 | -1.49 | 4.32 | 2.21 |
| | | (111.83) | (130.70) | (104.79) | (233.66) | (373.62) | (291.04) |
| **Lentil** | Export | 38.63 | -13.63 | 3.15 | -4.12 | 5.90 | 1.45 |
| | | (85.12) | (69.25) | (76.37) | (574.87) | (214.72) | (222.83) |
| | Import | 5.56 | 19.51 | 19.76 | 0.79 | 5.98 | 2.89 |
| | | (103.11) | (80.34) | (58.54) | (688.23) | (309.35) | (318.74) |
| **Peas** | Export | 25.86 | 14.14 | 18.69 | -9.88 | 3.58 | -3.19 |
| | | (73.64) | (90.54) | (77.41) | (194.05) | (232.80) | (218.08) |
| | Import | -14.63 | 6.70 | 4.14 | 4.99 | 3.28 | 4.19 |
| | | (118.18) | (266.97) | (126.45) | (444.24) | (354.57) | (351.28) |

*Parenthesis in the indicates CV (%)*

**Table 4. Growth rate and instability of export import quantity and price of oilseeds (1990-2016)**

| Crops | Variables | Quantity growth rate (%) | | | Price growth rate (%) | | |
|---|---|---|---|---|---|---|---|
| | | **1990-91 to 2000-01** | **2001-02 to 2015-16** | **1990-91 to 2015-16** | **1990-91 to 2000-01** | **2001-02 to 2015-16** | **1990-91 to 2015-16** |
| **Groundnut** | Export | 15.90 | 11.04 | 12.23 | 2.77 | 6.19 | 4.41 |
| | | (125.12) | (161.97) | (116.91) | (372.09) | (319.98) | (259.20) |
| | Import | -5.76 | 20.41 | 2.05 | 0.00 | 0.00 | 0.00 |
| | | (78.73) | (70.19) | (52.04) | (31.62) | (79.62) | (56.04) |
| **Soybean** | Export | 63.03 | 42.57 | 27.34 | -2.87 | -3.76 | 3.36 |
| | | (43.92) | (77.72) | (63.17) | (67.52) | (208.22) | (123.78) |
| | Import | -66.88 | 134.30 | -1.66 | 0.00 | -12.52 | 4.19 |
| | | (38.23) | (41.84) | (39.67) | (39.40) | (101.02) | (58.69) |
| **Niger seeds** | Export | 10.99 | -2.98 | 1.59 | -0.94 | 7.50 | 2.72 |
| | | (236.04) | (266.68) | (232.97) | (848.08) | (288.62) | (320.00) |
| | Import | 0.00 | 0.00 | 20.24 | 0.00 | 0.00 | 0.00 |
| | | (0.00) | (0.00) | (68.89) | (0.00) | (0.00) | (79.94) |

| | | | | | | | |
|---|---|---|---|---|---|---|---|
| **Mustard** | Export | 58.86 | 3.63 | 24.35 | -3.63 | 3.14 | 0.12 |
| | | (61.76) | (187.86) | (99.90) | (476.88) | (556.73) | (475.79) |
| | Import | 33.66 | -5.77 | 11.05 | 33.66 | -5.77 | 11.05 |
| | | (149.87) | (99.50) | (122.28) | (104.17) | (220.68) | (118.46) |
| **Safflower** | Export | -1.69 | 2.24 | 11.94 | 1.72 | 1.33 | 1.31 |
| | | (102.55) | (605.95) | (153.41) | (522.05) | (122.79) | (170.42) |
| | Import | 0.00 | 0.00 | 25.26 | 0.00 | 0.00 | 0.00 |
| | | (0.00) | (0.00) | (63.65) | (0.00) | (0.00) | (26.73) |
| **Sesamum** | Export | 20.83 | 2.74 | 10.79 | -2.70 | 6.58 | 2.01 |
| | | (165.01) | (301.91) | (156.38) | (625.95) | (250.03) | (218.10) |
| | Import | -10.46 | 81.45 | 21.22 | 0.00 | 0.00 | 2.37 |
| | | (93.49) | (72.17) | (53.65) | (69.92) | (264.14) | (122.06) |
| **Sunflower** | Export | -3.87 | 9.31 | 3.51 | 20.42 | 1.84 | 7.85 |
| | | (174.22) | (220.19) | (150.57) | (157.68) | (373.38) | (227.66) |
| | Import | -100.00 | 0.00 | 28.90 | 0.00 | 0.00 | 0.00 |
| | | (35.25) | (66.48) | (50.85) | (30.15) | (53.71) | (43.65) |

*Parenthesis in the indicates CV (%)*

### 3.3. Trade responsive of India's cereals, pulses, and oilseeds

How responsive are export and import quantities to a change in international prices is of direct relevance in international economics. Trade price elasticities are signals for exporters or importers to increase or decrease their trade as it indicates exports or imports responsiveness to changes in price.

Export import price elasticities of India's cereals, pulses, and oilseeds are presented in tables 5,6,7 respectively.During the period 1990-91 to 2015-16, the export import price elasticities of all cereals were found to be positive except export price elasticity of wheat (-0.30 %) crop. The export price elasticities of rice are found to be more than imports followed by maize and total cereals. During the same period, all pulses have witnessed positive and more export price elasticity than compared to imports. Among the pulses, more export elasticity was noticed in peas (2.36%) followed by gram, lentil and pigeon pea. During the period 1990-91 to 2015-16, the export import price elasticities of all oilseeds were found to be positive except imports price elasticity of soybean (-0.45 %) crop. The exports price elasticity of soybean (0.42 %), mustard (0.97 %), safflower (0.72 %), sesamum (0.58 %) and sunflower (0.09 %) were found to be marginally higher than the imports.

The study found that among cereals, pulses and oilseeds, rice (1.24 %), peas (2.36 %) and mustard (0.97 %) have high export elasticities respectively.

**Table 5. Export import price elasticity of cereals (1990-2016)**

| Crops | Variables | Export Import Price elasticity (%) | | |
|---|---|---|---|---|
| | | 1990-91 to 2000-01 | 2001-02 to 2015-16 | 1990-91 to 2015-16 |
| **Rice** | Export | 1.84 | 0.48 | 1.24 |
| | Import | 0.97 | 1.92 | 1.03 |
| **Wheat** | Export | 1.48 | -0.58 | -0.30 |
| | Import | 0.98 | 0.67 | 0.93 |
| **Maize** | Export | 1.40 | 0.68 | 0.93 |
| | Import | 1.26 | 0.67 | 0.57 |
| **Total Cereals** | Export | 1.69 | 0.26 | 0.91 |
| | Import | 0.82 | 0.56 | 1.03 |

*Source: CMIE Commodities, Accessed on 3rd June 2017*

**Table 6. Export-import price elasticity of pulses crops (1990-2016)**

| Crops | Variables | Export import price elasticity | | |
|---|---|---|---|---|
| | | 1990-91 to 2000-01 | 2001-02 to 2015-16 | 1990-91 to 2015-16 |
| **Pigeon Pea** | Export | 1.71 | -0.98 | 0.35 |
| | Import | 1.22 | 0.08 | 0.31 |
| **Gram** | Export | -2.33 | 0.37 | 0.69 |
| | Import | 0.91 | 0.36 | 0.53 |
| **Lentil** | Export | 1.61 | 1.21 | 0.55 |
| | Import | 0.83 | 0.40 | 0.47 |
| **Peas** | Export | 3.85 | 0.55 | 2.36 |
| | Import | 1.18 | 0.50 | 0.25 |

*Source: CMIE Commodities, Accessed on 3rd June 2017*

**Table 7. Export-import price elasticity of oilseed crops (1990-2016)**

| Crops | Variables | Export import price elasticity (%) | | |
|---|---|---|---|---|
| | | 1990-91 to 2000-01 | 2001-02 to 2015-16 | 1990-91 to 2015-16 |
| **Groundnut** | Export | 0.70 | 0.35 | 0.31 |

| | | | | |
|---|---|---|---|---|
| | Import | 0.72 | 0.00 | 1.00 |
| **Soybean** | Export | 1.38 | 1.78 | 0.42 |
| | Import | 1.00 | 3.83 | -0.45 |
| **Niger seeds** | Export | 1.17 | -0.41 | 0.25 |
| | Import | 0.00 | 0.00 | 0.67 |
| **Mustard** | Export | 1.45 | 0.43 | 0.97 |
| | Import | 0.00 | 2.93 | 0.00 |
| **Safflower** | Export | 0.00 | 0.57 | 0.72 |
| | Import | 0.00 | 0.00 | 0.00 |
| **Sesamum** | Export | 1.42 | 0.17 | 0.58 |
| | Import | 0.70 | 0.00 | 0.54 |
| **Sunflower** | Export | -0.09 | 0.70 | 0.09 |
| | Import | 0.00 | 0.00 | 0.00 |

*Source: CMIE Commodities, Accessed on 3rd June 2017.*

### 4.The Effects of International trade on, production, prices and poverty

Regression analysis was performed to study the effects of different trade variables such as EX-IM prices, EX-IM quantities, and trade balance on production, total exports and total imports and poverty.OLS regression coefficients(with one-year lag) indicatethe degree of responsiveness to any changes formed due to changes in price, quantity or trade balance. These results are useful to planners, policy makers in designing policies.

The empirical results of the oilseeds ordinary least squares (OLS) regression coefficients signals are in Table 8. Export price coefficients are signals for exporters to increase or decrease their exports as it indicates exports responsiveness to changes in price. The results on relevant variables are encouraging with expected sign and magnitudes of coefficients.

**Table8. The Effect of Economic Market Coefficients on production, total exports, total imports and trade**of **different oilseeds**

| **Parameter Estimate** | | **Oilseeds** | | | |
|---|---|---|---|---|---|
| | | Groundnut | Linseed | Soybean | Sunflower |
| **Production** | **Intercept** | 6471583.639 | 362951.6 | 3012839 | 1054522 |
| | **Export Quantity** | 243.4088858 | 0 | -5.78346 | -81.7645 |
| | | 1037.845382 | 0 | 5.768482 | 208.2874 |

| | | | | | | |
|---|---|---|---|---|---|---|
| | | **Import Quantity** | 0 | -61.9163 | 19.93212 | 0 |
| | | | 0 | 19.66929 | 25.89577 | 0 |
| | | **Export Price** | 1750.360229 | -18.5834 | 10341.2* | 82.31728 |
| | | | 2639.91095 | 14.42879 | 2067.074 | 208.2874 |
| | | **Import Price** | -70.86852276 | -87.4584 | 2484.729** | 20.46515 |
| | | | 194.8347362 | 27.01663 | 945.3928 | 16.17247 |
| | | **Adjusted R2** | -0.17866544 | 0.576338 | 0.671386 | 0.175858 |
| | | **R Square** | 0.041129243 | 0.679978 | 0.760403 | 0.341111 |
| **Total Exports** | **Agri** | **Intercept** | -71439.58691 | 186434.9 | -53356.8 | 65461.7 |
| | | **Production** | 0.002495707 | -0.66307 | 0.005438 | -0.08673 |
| | | | 0.003355207 | 0.240611 | 0.003564 | 0.026512 |
| | | **Export Quantity** | -52.2365062* | 37.40182 | -0.26063 | 40.60212* |
| | | | 16.56422012 | 32.62378 | 0.780901 | 8.493975 |
| | | **Export Price** | 80.88622099*** | 20.2586 | 224.8121* | 42.69902*** |
| | | | 41.33727352 | 20.30119 | 56.3844 | 22.91621 |
| | | **Trade Balance** | 52.5493195* | 0.895254 | 0.491442 | -22.516 |
| | | | 16.57651159 | 10.96147 | 0.769584 | 8.115279 |
| | | **Adjusted R2** | 0.919211102 | 0.483872 | 0.835369 | 0.806422 |
| | | **R Square** | 0.932137326 | 0.566453 | 0.86171 | 0.837395 |
| **Total Imports** | **Agri** | **Intercept** | 3002.13899 | -100.151 | -32131.5 | 50788.32 |
| | | **Production** | -0.001001644 | -0.1207 | 0.008124* | -0.05175 |
| | | | 0.002395606 | 0.100484 | 0.001813 | 0.012645 |
| | | **Import Quantity** | -19.52491879 | 28.62331 | 0.532168 | 22.20582* |
| | | | 11.95894568 | 10.78634 | 0.495162 | 3.65614 |
| | | **Import Price** | 0.837284019 | 54.64402 | -12.7332 | 0.583541 |
| | | | 2.180988243 | 15.67428 | 13.10913 | 0.839834 |
| | | **Trade Balance** | 0.178648285 | 28.40766 | 0.260751 | 10.49781* |

| | | | | |
|---|---|---|---|---|
| | 0.018336244 | 10.1492 | 0.064752 | 2.268161 |
| **Adjusted R2** | 0.819432882 | 0.714558 | 0.694454 | 0.82123 |
| **R Square** | 0.848323621 | 0.760229 | 0.743341 | 0.849833 |

*significance at 1%, **significance at 5%, ***significance at 10%

The results show that export price, import price coefficients have a positive and significant impact on Soybean production indicating export prices are influencing production. In case of Groundnut, export quantity has a negative significant impact whereas export price and trade balance have positive significant impact on the total agri export. Soybean export price has positive and significant impact on total agri exports.The results show that Soybean production coefficient, Sunflower import quantity coefficients have positive and significant affecting on total imports.

**Table9. The Effect of Economic Market Coefficients on production, total exports, total imports of cereals**

| **Parameter Estimate** | | **Cereals** | | |
|---|---|---|---|---|
| | | Rice | Maize | Wheat |
| **Production** | **Intercept** | 108487215.3 | 6449023 | 49338772 |
| | **Export Quantity** | 2.946994717* | 1.332822** | 2.914071** |
| | | 0.60271898 | 0.610083 | 1.054748 |
| | **Import Quantity** | -140.244408 | 31.96291* | 1.172067 |
| | | 82.94686092 | 10.36001 | 1.362281 |
| | **Export Price** | 28621.89447* | 24137.25* | 72926.14* |
| | | 9277.400035 | 8641.182 | 20868.98 |
| | **Import Price** | 60.0846528 | 5824.345*** | 20942.9 |
| | | 4362.311109 | 3162.396 | 22350.81 |
| | **Adjusted R2** | 0.68439785 | 0.776851 | 0.558216 |
| | **R Square** | 0.732952027 | 0.811182 | 0.626183 |
| **Total Agri Exports** | **Intercept** | -226542.964 | -111138 | -299301 |
| | **Production** | 0.001323479* | 0.011458* | 0.004674* |
| | | 0.000446776 | 0.003013 | 0.000996 |
| | **Export Quantity** | 0.205737148 | 0.035276 | 0.009779 |

| | | | | |
|---|---|---|---|---|
| | | 0.173457652 | 0.174891 | 0.009268 |
| | **Export Price** | 106.8526535* | 25.7507 | 126.7867 |
| | | 20.34820804 | 129.3172 | 108.7034 |
| | **Trade Balance** | -0.18954132 | -0.02563 | -0.00135 |
| | | 0.173384932 | 0.173437 | 0.005901 |
| | **Adjusted R2** | 0.943291381 | 0.826604 | 0.801978 |
| | **R Square** | 0.95236476 | 0.854348 | 0.833662 |
| **Total Agri Imports** | **Intercept** | -155028.117 | -80300.6 | -161610 |
| | **Production** | 0.001185076* | 0.008058* | 0.002641* |
| | | 0.000332504 | 0.001633 | 0.000472 |
| | **Import Quantity** | 0.20647301 | 0.023937 | -0.00249 |
| | | 0.152162937 | 0.091423 | 0.004979 |
| | **Import Price** | 8.143654624 | 11.89793 | 37.96466 |
| | | 7.129475542 | 24.15987 | 54.727 |
| | **Trade Balance** | 0.005892563* | -0.00701 | -0.00115 |
| | | 0.001484789 | 0.004969 | 0.002782 |
| | **Adjusted R2** | 0.799165942 | 0.791033 | 0.716809 |
| | **R Square** | 0.831299392 | 0.824467 | 0.76212 |

*significance at 1%, **significance at 5%, ***significance at 10%

The empirical results of the cereals ordinary least squares (OLS) regression coefficients signals in Table 9.The results show that export quantity and price on rice;export quantity and price and import price onmaize; export quantity and price on wheat have a positive and significant effect on their production i.e. as export quantity, price and import price increases, production increases showing a positive impact. Rice production and export price,Maize production has significant effect on total agricultural exports. This implies as production of rice, maize and wheat can positively affect the Total Agri exports which can overall affect the GDP of the country in a positive manner. Thus increase in production and agri exports can directly affect the overall working of the trade a very positive manner. Rice production as well as trade balance, maize and wheat production have a positive impact on total Agri imports.

**Table 10. The Effect of Economic Market Coefficients on production, total exports, total imports and povertyof cereals, pulses and oilseeds**

| Parameter Estimate | | Cereals | Pulses | Oilseeds |
|---|---|---|---|---|
| Production | Intercept | 158056202.3 | 8278183 | 206615 |
| | Export Quantity | 4.594068639* | 5.007247*** | -1.73238 |
| | | (0.784287386) | (2.649309) | (0.555489) |
| | Import Quantity | 2.977714962 | -0.11784 | -0.50118 |
| | | (2.843917366) | (0.288134) | (0.300136) |
| | Export Price | 86461.32087* | 1517.296 | -42.8443 |
| | | (30077.36152) | (2368.138) | (24.42783) |
| | Import Price | -10888.1684 | 10942.93** | 1.024497 |
| | | (28822.62763) | (4789.171) | (0.744207) |
| | Adjusted R2 | 0.711563669 | 0.670187 | 0.624962 |
| | R Square | 0.755938489 | 0.720928 | 0.68266 |
| Total Agri Exports | Intercept | -219058.623 | -272379 | 9322.554 |
| | Production | 0.000811973* | 0.015909* | -0.54685 |
| | | (0.000205624) | (0.003294) | (0.386497) |
| | Export Quantity | 0.006942121** | 0.116992** | 1.697974 |
| | | (0.002998213) | (0.047341) | (1.278739) |
| | Export Price | 156.3285961* | 89.30692** | 149.5726* |
| | | (28.65267829) | (33.0441) | (52.75274) |
| | Trade Balance | 0.001168439 | -0.01948 | -1.44059 |
| | | (0.00259027) | (0.004797) | (0.594575) |
| | Adjusted R2 | 0.958208825 | 0.931284 | 0.715187 |
| | R Square | 0.964895413 | 0.942278 | 0.760757 |
| Total Agri Imports | Intercept | -163234.536 | -110845 | 128465.5 |
| | Production | 0.000881307* | 0.007731* | -0.65237 |
| | | (0.000201047) | (0.001802) | (0.201445) |

| | | | |
|---|---|---|---|
| **Import Quantity** | -0.00077834 | 0.004252 | -0.23379 |
| | (0.00322669) | (0.022819) | (0.650218) |
| **Import Price** | -15.6597 | 8.033751 | 0.484478 |
| | (26.91288) | (37.39908) | (0.748576) |
| **Trade Balance** | 0.001595831 | -0.01233 | -1.14033 |
| | (0.001102381) | (0.023858) | (0.607392) |
| **Adjusted R2** | 0.780934941 | 0.926224 | 0.672156 |
| **R Square** | 0.815985351 | 0.938028 | 0.724611 |

*significance at 1%, **significance at 5%, ***significance at 10%

The empirical results of the cereals, pulses, oilseeds ordinary least squares (OLS) regression coefficients signals in Table 10. The results show that on cereals production;export quantity on pulses; export quantity and import price have positive and significant impact. Total Agri EX-IM play a key role in the working of the global trade economy. Negative impact can be setback for a country. On total agri exports of cereals and pulses;production, export quantity and prices have positive and significant. Whereasfor oilseeds, export prices have positive and significant effect. In case of total Agri Importof cereals and pulses, production has significant impact.

### Table11. The Effect of Economic Market Coefficients signalson poverty

| Parameter Estimate | | Cereals | Pulses | Oilseeds |
|---|---|---|---|---|
| | Intercept | 86.75 | 69.45 | 22.02 |
| | Production | -2.04E-07 | -1.5E-06 | 0.000147* |
| | | (2.23E-08) | (4.78E-07) | (2.59E-05) |
| | Import price | 0.003134 | -0.00521 | -8E-05 |
| **Poverty** | | (0.00345) | (0.012884) | (0.000135) |
| | Export price | -0.01 | -0.005 | -0.00731 |
| | | (0.004) | (0.0054) | (0.00432) |
| | Trade balance | -3.19E | 2.65E-06 | 0.000141** |
| | | (1.313E-07) | (6.67E-07) | (5.37E-05) |

*significance at 1%, **significance at 5%, ***significance at 10%

**Trade on Poverty:**According to FAO, "In attempting to combat hunger and poverty, developing countries face the challenges of identifying specific agricultural and rural development needs and opportunities, and focusing investment in those areas where the greatest impact on food insecurity and poverty could be achieved."The upward pressure on food prices makes it particularly important to focus on the potential for increasing productivity. Productivity improvements—particularly in developing countries—have been found to be a powerful force for poverty reduction (Datt&Ravallion, 1998; de Janvry&Sadoulet, 2002)

Taking into account the crops such as cereals, pulses and oilseeds, table 11 shows how production, IM-EX price and trade balance affect poverty in India.The results show that production of cereals and pulses have a negative relationship with poverty, whereas import price for pulses and oilseeds also have negative relationship with poverty. Export price for all the 3 have a negative relationship with poverty. Trade balance is negative on cereals whereas positive on pulses and oilseeds.Thus, Production for cereals and pulses, import price for pulses and oilseeds, export price for all the 3 crops have adverse impact on Poverty.Only for oilseeds, production and trade balance have positive and significant effect on poverty.

## 5.CONCLUSION

Agriculture production, trade and poverty are interdepending on each other in its own way. The study found that all India cereals, pulses, oilseed crops have comparative advantage during liberalizationexcept rice, maize and mustard. The trade growth of Indian Agricultural commodities shows that during the period 1990-91 to 2015-16, wheat export price and rice import price growth rates were found to be more than other cereals export import prices. The imports price growth rate of all pulses were more, compared to exports price growth rate indicating India is importing at higher prices. It is noticed that among the oilseeds crops, mustard import price (11.05 %) and sunflower export price (7.85 %) growth rate was more, compared to other oilseeds export import prices. Whereas the export import quantity growth rate of all cereals, pulses, and oilseeds were found to be positive except rice imports (-14.82 %), wheat exports (-3.05 %) and soybean imports (-1.66 %), which were negative.

Export import prices of all cereals were unstable except maize import price. And export as well as import prices of all pulses were found to be unstable. It is observed that the import price of all oilseed crops were found to be stable except mustard and sesamum import prices. The study found that among cereals, pulses and oilseeds, rice (1.24 %), peas (2.36 %) and mustard (0.97 %) have high export elasticities indicating these crops highly responsive to change in international prices.

The effect of trade on poverty shows that, Production for cereals and pulses, import price for pulses and oilseeds, export price for all the 3 crops have adverse impact on Poverty. Only for oilseeds, production and trade balance have positive and significant effect on poverty.Thus import prices of cereals, quantity of imports for pulses and export prices of cereals, pulses and oilseeds need to be thoroughly examined and pre-requisite actions as well as policy measures are required. The study advocates that India should design self-reliance policy in pulse production as India is importing pulses at higher prices.India also should design new stable foreign price policy for cereals and pulses as export import prices of all cereals (except maize) and all pulses were found to be unstable.

Note: MB Dastagiri (2019),**Certificate of Honor award** for phenomenal and worthy keynote presentation on "International Trade Economic Analysis for Indian Agriculture during WTO Regime: Effects on Production, Exports, Imports and Poverty"at 4th International Conference on "Advances in Agriculture & Animal Sciences towards Global Food Security" organized by Endling Conferences, USA in association with Life Science Society of Hyderabad, August 28-30, 2019 at University of Hyderabad, India.

.

## References

[1] Aksoy, M.A., 2004. "Global Agricultural Trade Policies." In M.A. Aksoy and J. C. Beghin (eds.), Global Agricultural Trade and Developing Countries. Washington, DC: World Bank.

[2] Alkhteeb TT, Sultan SA (2015). „Determinants of India"s Agricultural Export", European Journal of Business and Management, 7(4):53-62.

[3] Caine S (1958). Comments on Nurske"s paper. Kyklos, Volume XI, pp. 187-193.

[4] Dastagiri [2017] found that there is a need that the developing countries must follow strategic action to meet the future demands (2025) of the wheat, maize, soybean, pig meat, poultry meat, sheep meat, beef and veal.

[5] Dastagiri, M.B. (2017) Global Agriculture: Vision and Approaches. European Scientific Journal, 13, 312-325.

[6] David Hallam (2003), 'Commodity Market Review 2003-2004', UNFAO Commodity and Trade division, 2003, pp:3-18.

[7] Dastagiri, M.B. and Anjani, S.V. (2018) The Political Economy of Global Agriculture: Effects on Agriculture, Farmers, Consumers and Economic Growth. European Scientific Journal, 14, 193-222.

[8] Dastagiri and Vajrala [2018, 16] reported that, since 2000 among WTO member countries, Russia, India, China and New Zealand have given more domestic support and Mexico, Chile, USA, Japan and OECD countries have given decreasing support to agricultural sector.

[9] Dominick Salvatore, International Economics, 8th ed. (Hoboken, N.J.: John Wiley & Sons, 2004), 15.

[10] Encyclopedia (2010), 'World price', The Great Soviet Encyclopedia, 3rd Edition (1970-1979). The Gale Group, Inc. Farmington Hills, USA.

[11] Financial Dictionary (2012), 'World price', Ferlex financial dictionary, Ferlex Inc., Huntingdon Valley, USA.

[12] Glezakos (1973) "Export instability and economic growth: a statistical verification", Economic Development and Cultural Change, volume 21:670-678.

[13] Gohin.A, Moschini.G "Evaluating the market and welfare impacts of agricultural policies in developed countries: Comparison of partial and general equilibrium measures" REVIEW OF AGRICULTURAL ECONOMICS, Vol.28, Issue: 2, Pages: 195-211, DOI: 10.1111/j.1467-9353.2006. 00281.x, Published: SUM 2006.

[14] https://siteresources.worldbank.org/INTPOVERTY/Resources/335642-1124115102975/1555199-1124115187705/ch2.pdf

[15] Jubair.BN, Alhiyali.ADK "An Economic Study of the Impact of Foreign Agricultural Trade and some Macroeconomic Variables on the Exchange Rate in Iraq using the Fmols Model for the period (1990 - 2015)" IRAQI JOURNAL OF AGRICULTURAL SCIENCES, Vol.49, Issue: 4, Pages: 541-550, Published: 2018.

[16] M. Aslam Chaudhary and Ashfaq A. Qaisrani "Trade instability, investment and economic growth in Pakistan" Pakistan Economic and Social Review, Vol. 40, No. 1 (Summer 2002), pp. 57-73.

[17] Nurkse R (1958) "Quest for a stabilization policy in primary producing countries" Kyklos, volume 2:141-154.
[18] Santeramo.FG, Guerrieri.V and Lamonaca.E "On the Evolution of Trade and Sanitary and Phytosanitary Standards: The Role of Trade Agreements" AGRICULTURE-BASEL, Vol.9, Issue: 1, Article Number: 2, DOI: 10.3390/agriculture9010002, Published: JAN 2019
[19] Sarah Joseph "The WTO, Poverty, and Development"DOI:10.1093/acprof:oso/9780199565894.003.0006.
[20] Satyanarayana.V, Wilson.WW, Johnson.DD, Dooley.FJ "Impacts of Trade and Agricultural Policies on the World Malt and Malting Barley Market" CANADIAN JOURNAL OF AGRICULTURAL ECONOMICS-REVUE CANADIENNE D AGROECONOMIE, Vol.46, Issue: 2, Pages: 127-147, DOI: 10.1111/j.1744-7976.1998.tb00359.x, Published: JUL 1998.
Singh, K. and Sangla, K. (2012) An Analysis of India's Exports (1991-2006). ZENITH International Journal of Business Economics & Management Research, 2, 79-108. http://zenithresearch.org.in/

[21] The World Bank IBRD IDA, The Role of Trade in Ending Povertyhttp://www.worldbank.org/en/topic/trade/publication/the-role-of-trade-in-ending-poverty
UNCTAD (2009) Impact of the Global Slowdown on India's Exports and Employment. http://www.unctad.org/en/PublicationsLibrary/webditctncd2009d1_en.pdf

UPENDER, M "LONG RUN EQUILIBRIUM BETWEEN INDIAS'S EXPORTS AND IMPORTS DURING 1949-50 -2004-05" Applied Econometrics and International Development, Vol.7-1 (2007).

[22] Wiggins.S "Trade Reform, Agriculture and Poverty in Developing Countries: A Review of the Empirical Evidence" AGRICULTURAL TRADE AND POVERTY MAKING: POLICY ANALYSIS COUNT, Book Group Author(s):OECD; OECD, Pages: 287-296, Published: 2003.
[23] William Krist, Chapter 3: Trade Agreements and Economic Theory, Wilson Center Home About https://www.wilsoncenter.org/chapter-3-trade-agreements-and-economic-theory.
[24] Winters. LA "The European Agricultural Trade Policies and Poverty" EUROPEAN REVIEW OF AGRICULTURAL ECONOMICS Vol.32, Issue: 3, Pages: 319-346, DOI: 10.1093/eurrag/jbi016, Published: Sep 1 2005.
[25] World Bank (2014-15), Povcal Net, from Global Monitoring Report 2014/15.
[26] World Bank Group and World Trade Organization, 2015.The Role of Trade in Ending Poverty. World Trade Organization: Geneva

Made in United States
Orlando, FL
04 November 2026

90201141R00017